Critical Thinking

How to Guide your Life with Good Decision Making and Problem Solving Skills

Contents

Introduction

Why is it that for some people, making decisions or solving problems is an exercise in frustration, while others seem to have a much easier time? What determines our capacity for being a good problem solver? Is being a critical thinker something that you're simply born with?

You could think of your brain as the "control center" of everything that goes on in your body and mind. It's not only in charge of your thoughts, but also your feelings, and the actions you take every day. Your brain determines the type of person you are and how you react to others and your environment. Because of this, it's important to learn how to fully optimize its performance and harness its immense potential so that works for us in critical times.

A lot of people are well aware of the benefits of exercise, and they strive to keep their bodies in good condition so that they can enjoy a better overall quality of life. On the other hand, we rarely pay attention to our brains, and we simply hope that it is doing a good enough job without our involvement.

In this book we'll delve into why some people have an easy time solving complex problems in their lives, and how we can get to that point ourselves. We'll discover how we can eventually become great critical thinkers ourselves, so that we then can enjoy a more fulfilling life by reducing the levels of stress associated with the decision making processes that go in our minds so that eventually learn how to make the right choices that will have the most positive impact in our lives.

No matter what age or condition you are currently in, it is very likely that you can start exercising your brain right now and improve your “relationship” with it so that it’s in top form for when you’ll need it the most.

Thank you for reading this book. I hope you enjoy the journey!

Chapter 1: What is critical thinking?

The concept of critical thinking is nothing new. It has been developing for over two thousand years now. The term itself has its origin in the mid/late twentieth century.

Critical thinking is one of the most important and sought after skills in most industries and there's a big reason why. In a few words, it is the ability to analyze the way we think and present evidence for our ideas, instead of simply using our personal reasoning as proof.

There are several benefits that we can gain from developing and mastering critical thinking skills. For example, we can get a better understanding and empathy for other points of view, be in better control of our learning process and being able to solve complex problems.

In order to develop good critical thinking skills, we must have a good base of communication and problem solving abilities and be fully committed to overcome not only our own ego but sociocentrism too. Having little regard for the interests and beliefs of others and assuming the superiority or rightness our social group are one of the biggest enemies of critical thinking.

Good critical thinkers realize that no matter how skilled they are as thinkers, it's always possible to improve their reasoning and that they're not immune to human irrationality, self-interest, uncritically accepted rules in society, biases and distortion. They also recognize that developing as a critical thinker is a complex process, and are fully committed to life-long practice in order to improve. They fully embrace the famous Socratic principle: 'The unexamined life is not worth living', because they are aware how a lack of examination promotes an unfair, dangerous and uncritical environment.

The ability to think rationally has important benefits in whatever we choose do in life. There are certain careers where it is almost an explicit requirement if one wishes to be successful, for example: in finance, education, law, and scientific research. However, to be able to think well and solve problems is a useful skill that isn't restricted to any particular career choice or industry. It is a great asset for doing better in almost every area of our lives.

We all think, as it is in our nature to do so. But a lot of our thinking, is simply left to itself; which means that it is often biased, distorted or prejudiced. It's often uninformed too. Yet, a lot of the satisfaction we get from life, and what we're able to make and produce is largely dependent on the quality of our thinking. Having poor critical thinking skills often means not only having less money, but a subpar quality of living too.

When we thinking critically about an idea or issue, we greatly improve the quality of our thoughts by taking charge of the structures inherent in thinking and letting intellectual standards lead the way.

A good critical thinker can:

-Raise important questions and problems, and communicate them in a very clear and precise manner.

-Identify and assess relevant information and be effective at interpreting it with abstract ideas.

-Always be open-minded when facing alternative ways of thinking.

It's hard to dispute the importance of critical thinking and the benefits it offers to our lives. Over the last 20 years, journalists have caught on an have praised the merits of critical thinking in their articles while many schools have embraced it by offering courses and programs that help their students increase their logical thinking and problem solving abilities.

Despite the increased use of the concept, it's not always fully understood, and there are several myths that make it even more difficult to see its life improving potential.

Let's take a look at some of the most prevalent myths that are attached to critical thinking:

-The benefits of critical thinking are mostly academic. This is completely false. There are numerous studies that show how most employers place a lot of value in critical thinking skills and regard it as one of the most important skills that many college graduates lack. To these employers, being able to think critically is equivalent to being able to solve complex problems and adapt to new situations quickly.

And the benefits don't stop at the career level; as we'll see, critical thinking can impact some of the most important areas of our day-to-day lives positively.

-Critical thinking often leads to criticism or disapproval. This is not always the case. Being able to think critically often means being open minded when considering the ideas of others. It indeed may lead to a negative appraisal of others' point of view, but it also demands the ability to be your own critic, which then encourages the analysis and synthesis information for a better appraisal of those ideas. In other words, thinking critically requires you to sometimes change your beliefs and adopt those of some else. It demands a close inspection of relevant facts and evidence. If that inquiry confirms someone else's' ideas or beliefs, then the critical thinker will be always be lead by reasoning and logic.

-**Critical thinking can interfere with creative thinking.** While it is true that there is a structure of thinking process, someone with great critical thinking skills will recognise that there is no single correct way of thinking. What tends to happen is that critical thinking often gets a bad press because it is confused with 'criticism' or in other words finding fault in the ideas of others or censoring, which of course is something that would affect creativity.

-Younger children shouldn't be taught critical thinking. While it is true that it is important for children to learn basic subjects before they can take on more complex ones, it is a misconception that we shouldn't be teaching young children how to think critically. Smart teachers will know the importance of developmentally appropriate lessons. Preschool children are already able to think critically. As thinkers, they even have certain advantages over adults because they approach ideas from a fresh point of view, and are rarely biased or encumbered by biases or preconceptions.

Now let's take a look at some of the less-obvious benefits of critical thinking:

-It enhances your language and presentation skills: being able to think clearly and systematically can make a dramatic difference in the way we express our thoughts.

When learning how to analyse the logical structure of our ideas, our comprehension skills will be directly enhanced.

-It is crucial in the new knowledge economy. Our global economy is being largely driven by technology, which means that one has to be able to deal with sudden changes in an efficient manner. Our global economy benefits flexible intellectual skills, and those that are able to analyse information and solve complex problems. Good critical thinking helps with this and it is key for success in the fast-changing environments we live in.

-It helps cultivate creativity. Being able to come up with creative solutions to simple and complex problems is not only about having new ideas. The ideas presented must be useful and directly relevant to the issue at hand. Critical thinking has an important role when reviewing new ideas, selecting the ones that will work the best, and modifying them when necessary.

-Improving our self-reflection skills. To live meaningful lives with good structure and balance, we need to know how to justify and reflect on our actions and values. Critical thinking helps us with the process of self-evaluation.

-It helps with academic performance. Critical thinking helps you think 'outside the box' and can help you develop a solid set of judgement and evaluation abilities that will help you enhance your academic performance. Students that are able to analyze and critique ideas are also adept at making connections across disciplines and see knowledge as something applicable to their daily lives. They also able to understand information on a deep and lasting level. Students that are able to think critically, also use a more 'self-taught' approach in their learning. They rely less on their teachers for instruction and guidance and become more self-directed.

-It helps you separate facts from opinions. It's no coincidence that good critical thinkers are the voices of reason in times of crisis or mass hysteria. They are able to examine problems from all sides and make rational conclusions while not letting their personal biases interfere. Because of their ability to consider all options, they usually present the best ways to solve a problem.

-You'll be less likely to fall for scams or tricks. A good critical thinker approaches most situations calmly with a healthy degree of scepticism. People who lack critical thinking skills are more likely to assume that everything that they hear is true, without considering the source or facts.

-It helps you separate thoughts from emotions. Critical thinkers know how to differentiate a rational thought from an emotional response that is largely based on personal bias. Emotion is the antithesis of logic. While emotional intelligence is important in many situations, being able to come to conclusions based on facts will usually prove to be more beneficial when solving critical issues.

-It may help you land a higher paying job. It is no secret that a lot of the highest paying jobs require solid critical thinking skills. Being able to generate and communicate effective thoughts and make complex decisions is highly valued by employers. In many high paying jobs, job interviewers will ask the applicants certain questions that will help them assess their critical thinking abilities.

-It helps you reach your goals quicker. Because critical thinkers consider a lot of options before they present a solution or make a decision and are aware that time is usually an important factor, they consider the fastest method to reach their goals. They become skilled at efficiency. With a critical thinking mindset, you'll know how to filter out irrelevant information that wastes time and resources. It helps you prioritise the things that are essential to the process.

Of course, there are several other benefits to being a critical thinker, far too many to list here. But the ones mentioned will probably interesting and motivating to most.

With carefully structured training, most people can become a more informed and rational decision maker and problem solver. In later chapters we'll learn how to do exactly this.

Chapter 2: The mind-set of a critical thinker

A great critical thinker's mind may seem difficult to understand and complex initially. However, their thought processes are built upon a set of uncomplicated thinking patters that connect information together from a rational perspective that is detached from emotions and biases.

The best critical thinkers see everything as a piece of a comprehensive canvas that needs to be understood as much as possible before any answers are revealed. This is what lets them make excellent decisions that further aid them in overcoming the difficult obstacles they face in their lives.

Having critical thinking skills alone is rarely enough. A critical thinking mind-set should always accompany the skills. Individuals that have both a great critical thinking mindset and skills will have no difficulty believing that they are able to solve any problem and that no challenge is impossible. Thanks to their mindset, they approach challenges with a healthy dose of optimism, determination and confidence to improve the situation they are facing.

This makes a stark contrast to a defeatist or pessimistic attitude that accepts failure as the most common consequence of a situation, and believes that there is little that can be done to improve a situation.

Because of this, any person that aspires to become a great critical thinker should be aware of which principles underlie a solid critical thinking mind-set.

-There are no problems, just exciting challenges. Being willing to solve complex problems is one of the core attributes of a great critical thinking mind-set. Individuals with this type of mind-set are able to respond to problems by keeping a constructive and positive view about change. Compare this to the average person's sometimes crippling fear of change. Good critical thinkers are able to find excitement in challenges and problems to a degree that they look forward to searching for ways to solve them. When they are facing a strong change, they see an opportunity instead of a threat. Their curiosity and interest are sparked even more when the challenge is apparently unsolvable and they find a way to devote time and energy to solve the important issues in their lives instead of ignoring or running away from them.

-Embracing courage and risk. If there's one thing that most often inhibits a critical thinking mind-set, it is definitively the fear of failure. A lot of people sometimes have great solutions in their heads, but they fear pointing out new methods of doing things or the flaws of others because they live in constant fear of retribution and avoid wanting to be the one who 'stirs the hornet nest' and is not a 'team player'. Some of the biggest companies in the world have been built upon taking risks at times where others would be too afraid of change. Remember that if you're not making at least one hard decision every week, you're probably not pushing yourself out of your comfort zone enough. The more you face making challenging decisions, the easier it will be for you to see such situations as normal parts of your everyday life that can be easily dealt with. In time, you'll probably surprise yourself when you even start looking forward to them.

-There's no room for excuses. Yes, it is much easier for us to find excuses for our lack of success and blame other people or external factors out of our control for it. The companies and people that have fully embraced learning from failure and not using excuses are the ones that have the ability to rebound from tough setbacks. When things aren't going well, they realize that it's their full responsibility to work in order to make them better.

-Not being afraid of acting with incomplete knowledge. While we should always strive to collect as much knowledge and facts as possible before trying to solve a challenge, the reality is that often there will be not enough information to support our decision making process or there may be a lack of awareness of what is really happening. When facing these situations, we must be ready to act with incomplete knowledge. We won't always have the luxury of being fully aware of what's happening and if we wait until we gather more information, we could miss important windows of opportunity that may not open again. Because of this, we need to learn how to use our intuition to make decisions and move forward. Learning how to make decisions when we have incomplete information is a skill that we're able to improve, and it's critical to do so. It's important to pay attention to short moments of first reactions, or impressions whenever we are facing complicated situations and issues.

-Never stop questioning and acquiring knowledge. As mentioned, those with great critical thinking mindsets often label problems simply as interesting challenges. They also see these challenges as great opportunities to learn and grow. When you have an intense curiosity about a situation, you'll want to stay well informed, ask questions to gain more clarity, and be open to learning about new ideas. As an example, some of the most successful companies in recent years have been those that have embraced the unavoidable presence of social media. This has created lots of opportunities and challenges for companies that are open to new ways of thinking. They have adapted their processes to include new ways to market their products and services and interact with their customers.

-**Always be moving forward.** Often times, people tend to hit a challenging roadblock and as a result, they stop trying to develop a new idea or way of doing things. They are quick to discard after receiving some criticism. Good critical thinkers are aware of the fact that new ideas and ways of doing things won't always be quickly embraced by others at first. In fact, they are more often criticized and even ridiculed in some cases. Some of the changes that have brought positive impact to our world where first ideas that were thought of as crazy. Once the idea starts getting new acceptance, things change, and the originator eventually is seen as revolutionary or a hero. In the corporate world, it is common to see individuals that often suggest great ideas, but when they're not acknowledged by others, they quickly withdraw from the meeting or discussion and never mention the idea or suggestion again. People with a critical thinking mind-set aren't afraid of pushing through walls when they are confident enough in their abilities. Remember that it sometimes may take several attempts for an innovation to be accepted by other. This requires not only courage, but patience.

-**Keep an open mind.** Paradigm shifts can be extremely challenging and sometimes the most challenging and complex problems will require a paradigm shift to be overcome. The issue is that most people aren't able to conceptualize and embrace a new way of thinking. Individuals with a critical thinking mindset are always open to new ways of doing things and fully embrace paradigm shifts. They are aware that there's several ways to look at a situation, and that there's also an infinite amount of options and perspectives available to them whenever they wish to use them. A flexible and fluid thought process is a requirement to gain new perspectives about circumstances.

-**Realistic optimism.** Being blindly optimistic can be just as dangerous as being a pessimist. Whenever you think about important challenges, the key is to remain realistically optimistic and always anticipate positive outcomes. There has been enough research that has shown that realistic optimism helps us be more proactive and productive when we are faced by the possibility of failure. This is even more crucial when we are in the position of leading and encouraging others.

-**Distinguishing between pieces of information.** The best critical thinkers are able to distinguish between pieces of info that seem to be similar on a surface level. They know that making assumptions about the facts might lead to misinterpreting their circumstances. Because of this, they always try to distinguish conflicting facts to get a better perspective of the circumstances they are dealing with. They know how to consciously distinguish a fact from an opinion and an idea from an assumption. This helps them better understand the situation from an unbiased point of view.

-**Identifying patterns and connections.** Whenever we process information through any of our senses, our brains automatically categorize it in a way that promotes a better sense of understanding. A good critical thinker knows how to identify a wider range of patterns within a piece of information. This is something that is rarely done by the majority of the population. Because of this, critical thinkers are able to make important connections and associations with existing information their brains have stored and organized and gather better insight that others are rarely aware of. Interestingly, this is one of the key traits that geniuses also have, and what makes them different from everybody else.

Chapter 3: Developing critical thinking skills

Thinking is something that comes naturally to our species. We don't have to do anything special to make it happen, it just does. In fact, it is much more difficult to stop the thinking process and clear our minds of thoughts than it is to start it. However, we have the choice to make our thinking happen in multiple ways. For instance, we can choose to think positively or negatively, or we could let our feelings guide or thoughts instead of using rational judgement; we can also think in multiple ways, such as analytically, scientifically and strategically.

For most of us, the ability to think critically is the one that will arguably impact our day to day lives the most. It can help us gain a broader perspective of the situations and circumstances we face in life. This way of thinking will stimulate new ideas that let us get to solutions that would have been very difficult to find via standard thinking patterns. It is not an ability that's only reserved for leaders, students and researchers.

In this chapter you'll learn a very simple step-by step process for improving your critical thinking that will help you overcome challenging situations in an efficient and effective manner. You will also be able to develop an independent confident way of thinking that is less vulnerable to information and opinions that can harm your wellbeing and quality of life.

Here's a few principles that we should always keep in mind:

- Knowledge is based on facts and it can only be acquired through thinking and reasoning.

Beliefs can't be knowledge. Beliefs are opinions disguised as facts.

Opinions are not based on facts or reason.

Thinking critically is a process that's based on the following: applying analysis, synthesized data and the thinker's ability to process and assess the information received.

- It's necessary to learn **how** to think so that you know **what** to think.

A critical thinker should never accept without questioning what another has to say or take it as a fact.

Learning how to think doesn't need repeated memorization.

To become a solid critical thinker, you must learn how to be skilled at gathering, analysing, synthetizing, reviewing and apply data.

- Critical thinking is an organized an often structured process used to judge an idea or situation.

It is separate from emotional attachments and because of this is unbiased.

To increase your effectiveness as a critical thinker you should have data and facts for when an argument is refuted.

- Critical thinking equals a search for meaning.

The meaning is for yourself in what the other person is saying or implying.

Why is logic so important in critical thinking?

Let's start by defining the word logic: it comes from the Greek 'logike', which refers to the science or art of using reason. By applying logic to your thinking process, you can evaluate arguments and help distinguish between good and poor reasoning or between truth and falsehood. By using logic, you can analyse the ideas or thoughts of others so that you can make good decisions and form sound beliefs in your life.

Here are the steps you can use to approach an idea, argument or a situation.

Step 1. Begin by paying attention to the important details.

One of the most important parts of the process and a great way to start is to learn which details matter the most. When we're exposed to a lot of information and differing opinions, it's not difficult to get lost and confused with the information. Because of this, we need to be able to identify which details matter and which don't. Start by paying attention to your intuition or 'gut feeling'. When something doesn't sound true, that should be your first alert sign. From this point, you can then look for other holes in an argument.

-Is someone benefitting from the statement?: Whenever you hear an opinion, it's a good idea to find out if there's someone that could be benefitting from stating it. There's often a good chance that the person making an argument does so for some reason. Of course, this doesn't necessarily mean that a person's motivations makes their opinion invalid, but it's wise to try to find out what they might gain.

-Pay attention to the source: if you notice that something sounds off, try to track down the source where it came from initially before forming any solid opinions.

-Watch out for misdirection: people making arguments say misleading things, and we can easily get carried away if we do not pay attention to the important details. Listening to debates, particularly science debates, can be a very useful way to train yourself to pay attention to those details. The more you practice this, the easier and more naturally will critical thinking come to you.

You don't become a great critical thinker just by training your brain. Being a great listener is also part of the equation. Being able to pay attention to words and phrases that usually accompany a weak argument is crucial. Qualifying statements tend to be common warning flags. Here's a few of these phrases:

-For the time being.

-Depending on the situation.

-I'm not saying.

-As far as I know.

-I want to say.

-Under normal circumstances.

-Under the following conditions.

These are the type of statements that usually signal untruth. They also alert you that it's time to pay more attention to what the other person is saying.

Step 2. Using past experiences and asking basic questions.

It's now time to consult with your past memories, experiences and understanding of a similar situation, idea or challenge that you have previously experienced. Recall how you dealt with the situation or idea by identifying exactly which approach you used. In this early stage, you are basically trying to bring forth helpful insights from past experiences into the present.

Most people probably believe that great critical thinkers ask very complex questions when they're trying to find a solution to a problem or when they're trying to separate truth from falsehood, but surprisingly, they often do quite the opposite. The better critical thinker you become, the more direct and fundamental your questions become. To improve your questioning process, a good idea is to break your questions down.

Imagine that you're facing a new challenge in your life and are not sure what approach to take. You could start by asking the following questions:

-What information from my past experiences about this problem do I already have?

-How is this information relevant to the current situation I'm facing?

-Why did this help me in the past?

-What is my objective and what am I trying to prove, disprove, find out, or support?

-Is there anything important that I might be overlooking?

Asking yourself these type of questions can help you get directly to the source of a problem. It's often a good idea to start out with simple questions before using a higher degree of complexity.

It's common for an explanation to become so complicated that even the original question gets lost. In order to avoid this, we should remember to go back to the basic questions you asked at first. It's a good idea to write down the answers to the above questions so that you can better remember them later on. Interestingly, some of the most impressive solutions to problems are incredible not because of their level of complexity, but for how elegantly simple they are.

Step 3. Be aware of your mental process.

People that consider themselves great critical thinkers often focus their analytical abilities externally, and arrogantly criticize others while rarely considering their own faults. Being a great critical thinker involves a great degree of self-reflection. One of the main things you should be paying attention to is your own thinking process. Identify where it started, what it currently looks like and the direction it's taking. Our minds can sort through a lot of information at incredible speeds, but sometimes we ignore important details along the way. When we face a problem or a new idea, it's easy to view it only from our perspective and overlook how others might see it.

Our brains are prone to use 'cognitive shortcuts' to understand what's going on, and these shortcuts help us to make it through the day with often great results. On the other hand, they also tend to form unreliable biases that can lead us to make wrong decisions or keep us close minded when hearing other's ideas.

No matter how good of a critical thinker you think you are, it's important to accept that we all have biases and, you should always keep an eye out for them. Try to get into the habit of constantly asking yourself what you're assuming and why. Ask yourself what you believe about the situation and why it's important to you. Afterwards, try to see if there are any assumptions you might be making about other's behaviours. There is a lot of literature on biases and how they operate that you can learn from. For instance, there has been extensive research done on how certain situational factors can have a big influence on our character traits. We tend to make different decisions depending on things such as our hunger and relaxation levels, colors we're seeing, etc. The simple act of reading about these biases will help you quickly adjust your perspective. One of your main goals should be to bring those biases to light so that they don't interfere with your thinking process.

Step 4. Understanding the idea or situation.

Your goal now is to strive for a thorough understanding of the current idea or situation that you're facing. It's important to gather thorough objective insights about these ideas or situations. An efficient way to achieve this is by organizing, comparing, and interpreting your quandary by using different several perspectives. This is where a great critical thinker is able to distinguish between facts/opinions, ideas/assumptions, and causes/effects. Ask yourself the following questions:

What is truly going on here?

How does the current situation compare to my past experiences and understanding?

How could I look at this situation through the other person's eyes?

At the same time, be alert to persuasive techniques. Remember that arguments try to persuade you to another person's point of view. Everything in an argument is trying to persuade you into thinking or believing something.

In an argument, the main idea is called the contention; this is the claim that the other person is making. You can better identify the contention by asking yourself, what does the other person want me believe or do?

When you're attempting to solve a problem or understand an idea, it's always helpful to learn from work that has been done in the same area or similar. There's no need to solve a problem from scratch when there's others that have already laid the groundwork.

It's important, however, to be able to evaluate this information critically or you could easily make the wrong decision or reach an incorrect conclusion. Whenever you're dealing with evidence, question who gathered it, how did they gather it and why?

As an example, imagine that you're looking at a study that shows the health benefits of milk. At first glance, the study appears to make sense, but then you learn that it was funded by the dairy industry. This doesn't mean that the study's results are automatically invalid, but you should certainly ask more questions when there's an evident conflict of interests.

Step 5. Application and synthesis.

At this point, your objective is to come up with possible solutions to the situation so that you can give a definitive answer later on. The easiest way to do this is by applying the knowledge and insights you've acquired in the previous steps.

At this point, you won't be making any definitive decisions, so it's best to simply explore the different possibilities you come up with. It's all about keeping a flexible approach so that you gain a better perspective about your circumstances. Ask yourself:

What if__?

How could I use ___?

What would happen if I ___?

What approach would best expand the possibilities here?

Once you've done this, you're ready to synthesize and compile the information in such a way that it will help you gain a unique insight into the possible solutions you have identified. This is the point where great critical thinkers start searching for patterns and connections between the pieces of information they've gathered so that they can get new insights that they might have missed before. Ask yourself the following questions:

If I could do__, what would I do?

How could I improve ___?

Out of the possible solutions I've come up with, which one seems to be have the best potential?

How could I test the consequences of these possible solutions?

Step 6. Evaluation

The final step of the process is to choose a possible solution or idea out of the ones you've come up and be prepared to defend and test the validity of it. There is rarely an idea or solution that will meet all of the criteria, but the best ones meet the most important criteria and have little negative consequences or impact on other factors.

Your critical thinking process shouldn't stop the moment you select a solution or idea and implement it. Instead, you should also include a follow-up that lets you evaluate the outcome. Sometimes, solutions do not work out as you envision them. There can be many reasons for this: failing to consider important factors, not enough information, not spotting biases or misperceptions, or any sudden changes out of your control that affected the problem or solution. We live in a constantly changing world, so there will rarely be a guaranteed answer for all situations.

Collect data and determine if the results were as you intended. Whenever there is an unexpected result or outcome, good critical thinkers go back to their thinking process and make adjustments as needed. In some cases the adjustments are minors, but sometimes an entirely new strategy will be needed.

Remember to practice constantly.

Just like any other skill, in order to get good ad critical thinking, it's important to be constantly practicing. It's not necessary to complicate yourself, a lot of the practice can be done inside your head.

Here's a few practical exercises and strategies you can use to improve your critical thinking skills.

-**Read about how great leaders solved challenges.** Reading the biographies of great leaders and historical figures that helped shaped our modern world is one of the best ways to get an insight on how to they managed to problems in the face of great adversity. Search for authors that will expose you to new concepts and ways of thinking so that you are challenged to think differently.

Always be looking for new problems that need solutions. In your day to day life, there will always be problems that are waiting for someone to show up and 'adopt' them. See if any family members or friends are facing a challenging situation and try to suggest them a few possible solutions. If you are in an organization, there surely are several problems that you can find and adopt to exercise your critical thinking skills.

-**Study competitors.** If you are in a position where you're dealing with competitors in a business, you can always attempt to determine which strategies they are using and how they are making money. Identify which customer groups they tend to focus on and why they are doing good or bad. Then do the same for your own business and try to find opportunities for doing better than your competitors. If you have colleagues or employees that are constantly facing customers, you can suggest them to do this exercise so that they gain a better understanding on the competition's strategies.

-**See things through other's eyes.** A great critical thinker always likes to think outside the box and knows the value of looking at a situation or a problem through another person's point of view. If you are stuck in a rut, challenge yourself to see familiar situations in new ways. Try to think as though you were someone else and consider things from different angles and perspectives. Whenever you're trying to solve an issue, make a list of your main concerns, what you need to solve it, who is able to help you and what are the possible challenges you might find. It's possible that you'll find new ways to reframe challenges as opportunities.

-**Don't waste time.** We all love to relax, but if you catch yourself indulging into too much of a good thing, it will be helpful if you start thinking about how you can use your time more constructively. Determine at what time you usually do your best thinking and see if you can learn new things then.

-**Foster your curiosity.** We are all naturally curious. Just take a look at a young child and watch how they are constantly asking questions to their parents. This helps them quickly gather information about their environment and learn at an accelerated pace. However, as we grow older, we tend to bury this natural curiosity and we even get in the habit of hiding the impulse to ask questions. Unfortunately, this is not the proper way to develop a critical mind. We can still easily train ourselves to foster our curiosity productively. It only takes a conscious effort to ask more questions about the things you see in your daily life.

-**Practice expressing yourself.** The process of thinking critically includes being able to express your ideas clearly. You might find it difficult not to argue every thought you do not agree with, but you'll probably have better results if you only focus on your main arguments. This will make it much easier for others to follow your ideas and way of thinking. If you're trying to explain an abstract idea to others, assume that they will not understand or that they will have a very hard time understanding. Give them examples and analogies whenever you can to make things easier.

-**Write down your successes and failures.** It's important to learn from our past successes and mistakes. The easiest way to do this is to write down key decisions and the outcomes so that you can reference these entries whenever you need them. By doing this, it is unlikely that you will repeat a mistake twice and you'll also be able to examine your assumptions and logic with great detail, so that you gain better insight into your own critical thinking and problem solving process. You'll be able to easily identify your strengths and weaknesses this way.

-**Take care of your mind and body.** There's no doubt that thinking critically comes easier when you take proper care of your mind and body. Doing physical activity every day is not only good for your body, but it will help lower your stress and increase your mental sharpness so that you have an easier time making decisions and solving problems. Try to do challenging work when you are the most alert and you have the most energy. Don't be afraid to reach out for help whenever necessary. If you feel that you need some assistance with either your mental or physical wellbeing, talk to a health counsellor.

With critical thinking, there's no such thing as complete mastery. The more knowledge you are able to attain, the better that you'll become at it. Yes, you'll be constantly 'thinking about thinking', but it will result in a confident brain that is able to form great arguments almost automatically and present solid ideas and solutions as well.

Chapter 4: How to make better decisions

There's no denying that just about everyone in the world has made their fair share of regrettable decisions throughout their lives. If someone tells you that they've never made a bad decision they are probably lying or plain and simply avoids making decisions at all costs. Making better decisions is connected to having a critical thinking mind, and it's also a skillset that is important to develop.

Even great leaders fail. Why do they fail when making decisions? It's likely because they made poor choice that eventually lead to bad decisions. And often they pile up a several layers of bad decisions upon bad decisions. Leadership and making decisions are strongly connected. A leader's decision making process can greatly hinder or make them. People that have risen to the top in a company often (although not always) do so thanks to their solid decision making skills. Unfortunately, while it may take years or even decades of great decision making to rise to the top, while it sometimes only takes a single bad decision to fall. Whether you like it or not, when you're in the position of leadership or in charge of others, you'll be only as good as your last decision.

While intuition is a valuable tool in many situations where there's a lack of information or understanding, they will only take you so far in life. We shouldn't be relying on them instead of a solid decision making framework, as it will result on falling to an act of oversight, manipulation or other negative influences.

So how can we make great decisions? The first and arguably most important part of the decision making process is learning how to synthetize the information that we deal with in our daily lives, while striving to make the best decisions possible in a timely manner. This may sound like a tough code to crack, but in reality we only need to know how to properly filter information and to understand that there exist a pyramid of knowledge. If we successfully develop a qualitative and qualitative process for synthetizing information, we'll dramatically improve our decision making process in a short period of time.

Let's take a look at the 'pyramid of knowledge'.

At the top of the pyramid, we would find:

Intuition. Intuition is based on experiences and emotions and they not always have solid analytical support. As we know, when there is limited information, intuition might be all we have to make a decision. Interestingly, sometimes when there are more refined analytics at our disposal, our instincts can let us know if the other inputs are being reasonable or biased. The important thing to remember is that intuitive decision making is something that we can refine and work on. Working on becoming more discerning is what will give us the best returns.

Data. Raw data can be comprised of statistics, disparate pieces of information, or random inputs that when looked at individually, hold very little value. It's not advised to draw conclusions based on raw data, as it will often lead to flawed decisions

Information. We could see information as a more complete set of raw data. It is usually derived from a collection of data that has been processed and where two important elements, context and meaning, have been added to the disparate facts. As a results, this allows for a much better analysis.

Knowledge. When information is refined via analysis, and has been assimilated, put to test and verified, it is considered knowledge. Knowledge has a high degree of accuracy, because there is evidence (typically derived form an experiment) to back it up.

Although a lot of people consider theories and opinion as fact, they are definitively not in the same category. We tend to make ill-advised decisions when they are done under a lot of pressure and when not using a solid decision making structure or framework. The advantage of intuition based decisions is that they can be made rather quickly, but they carry high levels of risk. If we move a step further and make a decision at the information level, we will have a better degree of risk management but it still isn't as safe as doing a decision made based on information that has been tested and verified.

Another important thing to consider when making decisions is the actual source of the input. Great leaders know when to seek the advice of others, but ultimately they have the final word and are in control of the final decision. However, an important thing to consider is where and how much advice you should seek. Incorrect input or simply more input won't add anything positive to your decision making process. Volume in the form of more input tends to make things more confusing, and seeking for advice from sources that don't offer any significant contributions is usually a big waste of time. Two other important factors that should always be taken in consideration when making decisions are the following, and they relate to the source of input:

-**The source's credibility.** It's important to consider the source's track record. What does deliver? Is it raw data, information or knowledge? Is the source only telling you what you wish to hear or does it give you the unadulterated truth?

-**Bias.** You should always strive to find out if there could be any hidden agendas that might be influencing the input received. Does the input provided benefit the source in any way?

Nowadays, with modern society's complexity, the ever increasing expectations of performance, coupled with the speed at which decisions must be made equals a lot of disastrous decisions. To counter this, we should use a guideline or structure when making decisions to minimize risk. By incorporating the following actions in your decision making process, you will diminish the possibilities of making bad decisions:

- **Analyze the cost-benefit ratio of your decision.** Important decisions usually have a cost. Try to think about what this is and find if the benefits justify it. Many bad decisions are made when the cost to benefit ratio isn't taken into consideration.

-Let others examine your decision. When making decisions that will affect others, remember that there is no such thing as a private decision. At some point, the details surrounding the decision you took will come out. A good test is to picture your decision printed on the front page of a newspaper or published on the front page of an important news site. How would this make you feel? If your family and friends knew about this decision, what would they think? When dealing with organizations such as companies, what would the shareholders and employees think? I possible, try to hear feedback form others before making a decision.

-Analyze the risk-reward ratio of your decision. Another thing to consider, is the reward to risk ratio. What are the rewards that may result from your decision and the potential risks? Are the odds against you or in your favour?

-**Ask yourself if it's the right thing to do.** It doesn't take much courage to stand behind a decision that is supported by the majority. On the other hand, being able to stand behind a decision that you believe in when facing disapproval and controversy is the stuff that legends and great leaders are made of. If you find yourself considering a decision that would yield big benefits but as a result it would compromise your integrity or values, you may need to thread carefully. Sometimes it's best to ask yourself if it's the 'right' thing to do. It's hard to go wrong by doing what's right.

-**Execution.** What good are all the previous steps if you fail to take action? Once you are sure about your decision, you should move forward and be firm. Do your best to learn how to make the best decisions possible even when you are dealing with incomplete sets of data. Sometimes, even with information gathered and the steps mentioned above, we tend to fall to analysis paralysis. Remember that windows of opportunity will rarely remain open for long periods of time, and the law of diminishing returns does apply here too; the longer you wait to take action and seize an opportunity, the more the returns will diminish. In many cases, the opportunity will completely disappear if you remain inactive for too long.

In addition to the above steps, you can also have a backup plan ready in case you end up making a bad decisions. Even when you've managed to gather knowledge from reputable sources and use a solid decision making framework, you're bound to make a bad decision. Great leaders know what to do the moment they realize they've made a bad decision. A great critical thinker is aware that all plans have constants and variables, and often times, the variables end up working against you. Good decision makers always have a backup plan ready in case that the circumstances fall beyond their reason or what they're able to control. Failing to have a backup plan equals leaving yourself at the mercy of variables.

Additional advice when making decisions

In addition to the previously mentioned framework, there are a few other things you should keep in mind that will help make the decision process easier and help you remain more calm in times of stress.

Be aware of your emotions. Many times, decisions are very personal and they are inevitably fuelled by our own feelings, outside influences and our relationships to others. Whenever you're making a decision, try to tap into your immediate feelings. When you are more aware of where your emotions are coming from when facing stress, you'll know if they could be influencing your decision or not.

One of the most powerful and influential emotions in decision making process is of course, fear. When making decisions, its best to remain realistically optimistic in order to avoid fear. When we approach a subject without taking the time to look into it deeply enough or considering all the options, we tend to have more regrets about decisions that go wrong.

-**The dangers of worry and stress.** Decision making tends to be especially complicated for people that suffer from an anxiety disorder, because they excessively stress about the possible negative outcomes when making a bad decision. This may happen even when making simpler decisions, for instance, when choosing a vacation spot. Excessive worrying about whether the place will have enough things to do or if it's too expensive or not might be unproductive and often end with the person not visiting any place at all.

-Think about the situation in a different language. If you're having trouble removing the emotional factor when making decisions, you may always try thinking about the situation in a foreign language. Doing this helps remove some of the emotional connection you might be using when thinking about things in your native tongue, as it will help you disconnect better from the situation.

How your environment affects your decisions. Recent studies have found that the physical location where you make a decision has an impact on your decision making process. Bright lights tend to intensify the initial emotional response we have to different types of stimulus. Whenever you need to make an important decision, a dimmer room works best to connect with your logical side.

Chapter 5: How to effectively solve problems

There's no denying that dealing with problems can be tough and sometimes makes us feel completely out of control in a situation and may even paralyse us.

Effective problem solving is part of a great critical thinker's skill set and it takes a degree of creativity. Unfortunately, most people don't consider themselves as the creative type and place themselves inside the 'I can't' box rather quickly.

As you already know, a great critical thinker sees problems as interesting challenges. The first step to become a great problem solver is to see problems as blank canvases waiting for answers that can improve your quality of life or that of others.

Whatever issue you might be dealing with, there are certain steps you can take to feel more in control. And even if you end up making a wrong decision in the end, you can learn how to feel confident with the actions and decisions you make.

Step 1. Define the issue.

The first step to solving any issue is to define it and look at it from multiple perspectives. If you are in a team or organization, what would your team mates, employees or customers identify as the problem? Issues tend to look different to different people.

When looking at a problem, also consider the relationship between cause and effect. We all have been in a situation where we solved symptoms and not the root cause of the problem. This means that the real issue doesn't go away and it simply appears as a new problem. If you take the time to look through different points of view and found root causes, you'll end up with a very clearly defined issue that you can now confidently approach.

-Step 2. If necessary, break it down into smaller pieces.

Sometimes, problems seem too big to tackle. If so, try breaking it down into small subcomponents. Smaller problems feel less overwhelming and they let you focus on details that often get lost when you are studying the full problem. For instance, if you are managing a company and you're having problems getting enough clients, break down the issue further and find out how much effort is being put into sales and marketing. You then might explore what is going on with the sales and marketing departments and your odds of finding the true root of the issue dramatically increase. This is also a great way to add some structure to your problem solving process so that you can be more direct in your research and analysis.

-Step 3. Generating and proving hypothesis.

Once that you've clearly defined the issue (or issues if you break it down into subcomponents), then it's time to start thinking about ways to solve it. Now is not the time to put them to action but simply come up possible solutions for each problem. See these solutions as the hypothesis that you'll need to prioritize, analyze and review. Continuing with the previous example, if the issue was not getting enough clients, some possible solutions would be to expand your sales team or change your current marketing strategy or perhaps increase your social media presence. These would be the initial three hypothesis for the problem.

Once you've come up with a list of possible solutions, it's time to prioritize your efforts. A big mistake is to waste lots of time by simply trying to prove every solution you come up with instead of using most of your time, energy and resources on the ones that have the potential for best return on investment. Make a reality check and evaluate the list of possibilities so that you can rule out the ones that seem the most unrealistic or unhelpful. Go through them and write a list of their pros and cons to determine this.

Step 4. Implementing the solution.

Once you have identified the solution that seems to give the best return on investment, it's time to do some planning and execution. First think about all the necessary steps that you'll need to take. Be as detailed as possible so that you don't miss any important piece that might affect the outcome. Then, it's time to take action. Be confident in the solution you chose and stay committed to id.

Step 5. Review the results

Review what you did. To know if you successfully solved the issue, it's crucial to review what worked and what did not, and what final impact the solution had. The review process will help you improve your long term problem solving skills and keep you from using extra resources and energy when not necessary.

Here's some extra advice that when practiced will quickly cause a positive influence on the effectiveness of your decision making process:

-**Practice brainstorming often.** Creativity is a key element of good problem solving. Even if you feel like your creativity is severely underdeveloped, it is possible to train and build it just like you would do with your body through physical exercise. Try to get into the habit of having regular brainstorming sessions. Think of a challenging situation that you've recently faced in your life where you haven't found a solution yet. Then write down as many what ifs? or open questions as possible. Something important is to remain realistically optimist while doing this. If you find that at the end of the session you came up with too many ideas, it's important to filter and connect the best ideas. This can be practiced by yourself or in a group setting.

-**Eliminate doubt.** When you're facing complicated issues and are having a hard time finding an obvious answer, the first thing that might pop in your mind is 'I can't'. If you ever find yourself in the 'I can't' box, then ask yourself why is it that you feel stuck. Questioning the reason why you feel unable to act can help you get to the root cause of a problem.

Thinking that you can't do something is often times an irrational belief and asking yourself exactly why you can't sheds light on it.

-**Change the way you see the problem.** Shifting the way that you see a problem is a very efficient strategy, because problems are usually perceived to be more challenging than they really are, causing a great deal of intimidation and analysis paralysis. Be aware of this intimidation factor, and instead of inviting anxiety and letting fear influence you, see problems as invitations or challenges to keep asking questions.

What to do when you're unable to fix the problem

Despite our best efforts, there will be times when we are unable to fix something, we might have tried a few possibilities and haven't had any success. If this happens, we might need to either focus on our coping skills or taking some time off before approaching the problem again.

If you find that you're experiencing several negative feelings because of the problem, it's important not to forget about yourself. Take some time to do things that you enjoy and relax you. It may also be helpful to find someone you trust that can give you good moral support. If you find that the situation is interfering with your life, it's probably time to seek out professional help.

Making decisions in a group or team setting

Great critical thinking skills are very important for effective teams and groups. Good teamwork requires members of the group to have several qualities, such as being skilled in effective communication and having the same goal in mind.

Group decision making is a process in which several individuals act collectively to analyze situations and come up with the best solution when selecting among the alternatives. When the team or group members possess good critical thinking skills, great results tend to occur. These skills prompt individuals to ask the right questions and give their opinion about situations and problems.

Decision making in teams or group may be informal in nature or be specifically created to work towards a specific common goal. The members of the group or team may be demographically similar or very diverse. Factors such as composition of the group, number of people involved, structure and purpose will all have an impact on the group's functioning to some level. Other factors such as time limits and conflicting interests can also impact the effectiveness of team or group based decision making.

Although many use the words team and group interchangeably, we should know how to distinguish both:

-Groups usually have a clear leader. In a team setting, there can be shared leadership roles.

-In a group setting, members usually have individual accountability while in a team there can be both individual and collective accountability.

-In a team setting, performance is measured through the collective result, while in the group, effectiveness is measured indirectly.

There are several effective methods and processes that can be used by groups and teams for making decisions. These are designed to make the decision making process better in some way. Here are some of the most effective:

-Brainstorming. Brainstorming sessions are typically unstructured to some degree and it involves the members of the group or team suggesting ideas or efficient courses of action to take. The situation or issue is described in as much detail as possible, so that every member has a complete understanding of it. Then the leader or the facilitator (a group member that helps understand the situation or common objective without taking a particular position) will ask for ideas to the other members. It is important to make a distinction between this stage and the evaluation stage, as the members shouldn't being evaluating the suggestions until all the ideas have been presented. Once no more ideas are presented, the members start evaluating the different options presented.

A common difficulty when brainstorming in a group or team setting is that some members will not present their ideas because of fear for being judged by others. To help with this, the leader or facilitator may propose an anonymous process for presenting ideas, such as writing them in paper or sending them via anonymous e-mails.

-**Dialetical inquiry.** This method focuses on the full consideration of the alternatives proposed either by brainstorming or any other process used. The leader or facilitator divides the team or group into opposing side, in which they debate the different pros and cons of the proposed ideas and solutions. Another alternative to dialectical inquiry is to use devil's advocacy, where a group member highlights the potential problems of the proposed ideas. These two methods make sure that the whole group considers all the possible consequences of their decisions.

-**The nominal group technique.** With this method, the decision making process is structured in such a way where group or team members have to compose a list ideas in writing. These ideas should be recorded privately. Once this stage is finished, the leader or facilitator asks the members to provide one idea from their list until all the ideas have been recorded on a board or chart for everyone to see. The members then engage in a discussion of the listed alternatives, on which they may give each of them a rating or ranking.

Delphi technique. In some instances, group members will be in different physical locations. The Delphi technique can be extremely useful in this case. It is primarily about harnessing and organizing complex issues that may require intuitive interpretation or informed guesswork of a situation. This technique relies on a panel of experts. It's also most popular with bigger groups and organizations. Experts are chosen due to their relevant expertise on the topic being examined. These experts may be on your project team, or may be other experts from the same organization or industry. Once the participants are confirmed, questionnaires are sent out to each member with instructions to comment on each topic based on their experience, research or personal opinion. The questionnaires are then returned to the leader or facilitator who then groups the information and sends a copy to each participant (and each participant has the option to comment further).

At the end of each comment 'round', the questionnaires are returned to the leader or facilitator, who will then decide if another round will be necessary of if the results are ready to be shown. These rounds can be repeated indefinitely until a general sense of consensus is achieved.

Pros and cons of group decision making.

We cannot conclude that group decision making is always better or worse than individual decision making, because its effectiveness depends of several factors. For example, demographic diversity can have an effect on the effectiveness of the group or team's functioning. Demographically diverse groups sometimes have social barriers to overcame, which may slow down the group. On the other hand, when properly managed, this diversity tends to generate a wider variety of options and better quality of decisions than more homogenous groups.

While we have to consider that situational factors do impact the effectiveness of group decision making, there has been enough research done through the years that offers general guidance about the pros and cons that are inherent to this type of setting. Here's some of the biggest advantages and disadvantages:

Pros.

Good group and team decision making involves making the most out of the diverse sets of strengths, experience and knowledge of its member. By using the unique qualities that each bring to the table, it is likely that in a group setting a higher number of alternatives that are of better quality than the individual can be generated. Because of this, the group or team may eventually reach a better quality solution than the individual.

Group decision making also has another important benefit: it often leads to a better collective understanding of the course of action chosen, since in most group decision taking methods, many members participated to arrive at the final decision. This helps promote a sense of acceptance of the decision, which in turn has a positive impact on the commitment of the individuals. It's easier to make the course of action successful this way.

Cons

Because groups are usually slower to arrive at conclusions than individuals, it can be difficult to use group or team decisions in situations where time is critical. The phenomenon of 'groupthink' is also an issue. In groups, pressure can cause certain things such as the deterioration of mental efficiency. In many cases, members will choose to set aside their personal beliefs and prefer to keep the peace rather than to go against what others are saying. Because this, there are many alternative courses of action that are never explored.

There are several reasons why groupthink occurs. For instance, if the group or team doesn't have a structured process for providing ideas and evaluating them, it is possible that many alternatives won't be explored and as a result the course of action taken might not be ideal.

The methods we explored in this chapter usually help reduce groupthink by making sure that each member offers something and considers the alternatives.

Another common disadvantage of group decision making is group polarization. This is when groups end up selecting more extreme solutions to a problem. Because accountability and responsibility is split, group members will sometimes go for riskier decisions than those they would go for if deciding as an individual.

Many individuals have to accept the fact that group decision making is an important part of their organizational lives. Because many spend some time in decision making groups, there are lots of research being done on this subject every year. Despite this, there is still a lot of things we do not know about the development and decision making processes of groups. Some of the current research is focused on identifying how factors such as cognitive or demographic diversity affect the final results. It is also likely that we will have a better understanding of how external circumstances affect their functioning.

Conclusion

As you know by now, critical thinking is one of the top skills that most successful people and high achievers learn to master. Its elusiveness as a trait could be attributed to the fact that it requires a very strange blend of two attributes that are rarely found on a single individual: humility and confidence.

Remember that, just like any other skill, developing a critical thinking mind is not an easy process. It is supposed to require a great deal of effort; if this were not the case, successful people would be way more commonplace in our world. Although not an easy path, the benefits that you'll be able to enjoy can be astounding. You'll not only make better decisions in the big things in life such as relationships, career and health, but also in the smaller ones such as navigating in a foreign city or choosing your next vacation.

The true power of critical thinking is that it will provide you with a great deal of self-reliance which will include empowered decision making and a high degree of confidence in your life

We all feel discouraged at the beginning when our small actions seem to have little effect, but remember that every great change starts with a tiny step. Every significant accomplishment owes a lot to the foundation that came before it.

Finally, if you enjoyed this book, please consider leaving a review on the platform. Reviews are one of the easiest ways to support the work of independent authors.

Thank you, and good luck!

Printed in June 2019
by Rotomail Italia S.p.A., Vignate (MI) - Italy